Vision
of the Honeybee Mimic

Adrian Horridge

Vision of the Honeybee Mimic

Published in the United Kingdom by
Northern Bee Books,
Scout Bottom Farm,
Mytholmroyd,
West Yorkshire HX7 5JS
Tel: 01422 882751
Fax: 01422 886157

www.northernbeebooks.co.uk

ISBN 978-1-914934-75-9

Design and artwork, DM Design and Print

Cover: courtesy of iStock

Nothing to worry bee keepers; not a pest; just a passing friend.

Dronefly, *Eristalis tenax* Note the large eye, compared with the honeybee.
US Dept of Agriculture. Photo by Stephen Cresswell

Among the fly family Syrphidae are many examples of bee mimics, mostly of the genus *Eristalis*, among which we found the world-wide common dronefly, *Eristalis tenax*, most convenient for detailed study, as there is no sting. As would be expected for a fly, the eye is much larger than that of the honeybee. Our analysis revealed it as a typical fly visual system with some specialised differences related to detection of movement and foraging among flowers for nectar. *Eristalis* does not mimic the honeybee in any way except appearance. The life histories are totally different. *Eristalis* has an aquatic larva that lives in nutritious rubbish, in which it has travelled independently around the world in bilge water of ships.

Female Eristalis

Male Eristalis

Contents List

What It's All About; and My Purpose

The large populations and short life cycles of most small invertebrates allows for the rapid evolution which is required to generate a mimic. Every step along the way helps survival, or it would not get built in, and reproduce successfully. Secondly, models for aspiring mimics to copy are abundant, often with a sting at one end or fierce jaws at the other. So, it is not surprising to find that mimics of bees or wasps are quite common. They must have evolved recently, because flowering plants offering nectar for bees, and bee habits, have all evolved together only recently.

My intention is to show that a mimic is not only skin deep. The behaviour must also mimic, and therefore the nervous system must also adapt. The finest details of the neurons in the nervous system can be described, and the activity of nerve cells in the visual system of the bee model and of the mimic can be recorded with microelectrodes, showing that mimicry goes deep into the structure and activity of the eyes and brain.

Fly Vision; for Escape, Colour and Flight Control

At first consideration, the vision of a typical fly, like the housefly, is quite unlike honeybee vision. Bees clearly use vision to recognise familiar foraging locations or to return to their hive. Flies have no such attachments, and poor memory, but they fly repeatedly to places and odours that attract them. Bees are always occupied in finding pollen and nectar using memory and odours, but house flies spend long periods wasting time, or flying round a central light bulb near the ceiling, which has the practical purpose of causing an upward current of warmed air that brings odours of food to them. Other common flies, *Lucilia* feed on sheep blood, *Sarcophaga* on carcases, and the bee mimic *Eristalis* visits flowers for nectar. All rely upon odours not vision to find food.

The Priority for a Fly is a Rapid Escape

Have you ever wondered why you can never hit a fly that settles near you, or if a fly sees colour, like a bee? Maybe fly vision is quite different from that of a bee; maybe not; we shall see.

First, the superb speed of the escape response of the fly is mostly due to the very short distance that the nerve signals travel between the visual receptor cells and the muscles of the legs that get the fly airborne. In humans, neuron distances are more than a metre altogether. Secondly, the neural circuit has been simplified by omitting the detection of the motion of your hand. The fly responds to the sudden *decrease in intensity at the receptors*, i.e., a response which is completed in 5 msec, omitting the processing of hand motion, which takes up to 50 msec (Holmqvist and Srinivasan, (1991). Having adopted the shadow as the signal, further selection increased the strength of this signal in a variety of ways; see below. Thirdly, even if you hit the airborne fly, it is not harmed because a fly has negligible inertia and is flexible, so it bounces off your hand

The very fast reaction of the fly has evolved to escape the vast number of their natural enemies. Millions of spiders in webs and armies of jumping spiders depend on flies for food. In the air, swallows collect mouthfuls of flies to feed their young. Praying mantids and frogs are specially adapted to catch flies. Fortunately for them, flies breed and mature quickly, and the larvae feed on corpses or rotting vegetation, which is abundant everywhere, Moreover, flies have superb sense of smell, and with seek out the slightest odour of food or water.

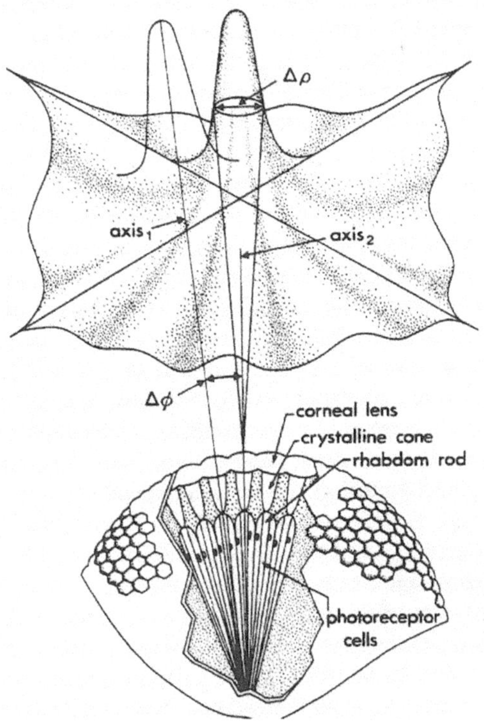

Figure 1. (Lower diagram). In most large day-flying insects, the rhabdomeres that absorb the incoming light are fused to form a central rod with one optical axis in each ommatidium. This arrangement divides the light coming through each facet into eight parts, reducing sensitivity, but with the advantage that each cell absorbs its own fraction of the colour mix, and the separation of colours is optimum. (Upper figure).The fields of view each have an angular width $\Delta\rho$, and adjacent facets overlap.

The Ommatidia of Other Common Insects

In the compound eye of a locust, butterfly or honeybee, each facet contains 8 or 9 long and narrow photoreceptor cells that contain the visual pigment, which happens to be like our own, rhodopsin, in an organelle called the rhabdomere (Figure 1). The rhabdomeres are close together, forming a rod, called the rhabdom, that carries the light and absorbs it at the same time. This means that each ommatidium has a single field of view, and the light entering each facet is divided between eight receptor cells. For optimum resolution and sensitivity, the fields of view of neighbouring ommatidia overlap (Figure 1, top).

Fly Ommatidia Collect More Light

Beneath the lens in each fly ommatidium there are six cells with separate axes, surrounding two smaller cells, one on top of the other, that share a common optical axis. As seen in a transverse section across the ommatidium, they form a pattern that is a mirror image in the two eyes, and in the upper and lower halves (Figure 2A).

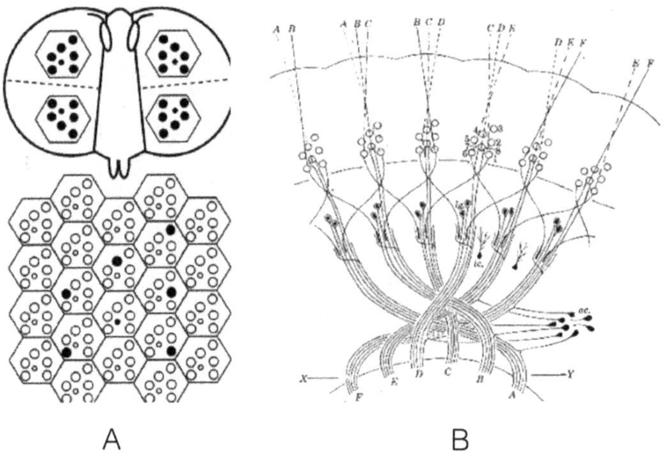

Figure 2, A The upper part of this figure shows the pattern of seven visual axes in ommatidia in different parts of the eye, seen from above. The lower part shows how the seven receptor axes from one ommatidium are distributed at the first synapses, in the second layer, the lamina. B. A vertical section showing in succession the optics, the retinal receptors, the synapses upon the lamina cells (shown in black), and the cross-over from front to back before the next region, the medulla.

This gives each receptor in the ommatidium a different field of view, but there are seven ommatidia that have the same field of view, and on their projection to the lamina below, the axons spread sideways to recreate proper projection, but now there is summation of seven cells at each visual axis in the lamina (Figure 2B). Here the seven retinula axons terminals that look *in the same direction* form a succession of large synapses upon the axons of the large monopolar neurons of the lamina (Figure 2B). The aggregation of seven receptor cells, and large number of synapses formed by each retina cell, amplifies the signal and speeds up the initial rise in the response of its lamina cell.

How Maximum Speed of Take-Off is Achieved

From the lamina onwards through the nervous system to the motor neurones to the leg muscles, the excitation is carried in axons that are longer and thicker and therefore faster than other neurons. In all parts of the insect nervous system, we find this difference between thicker faster neurons and thinner slow ones.

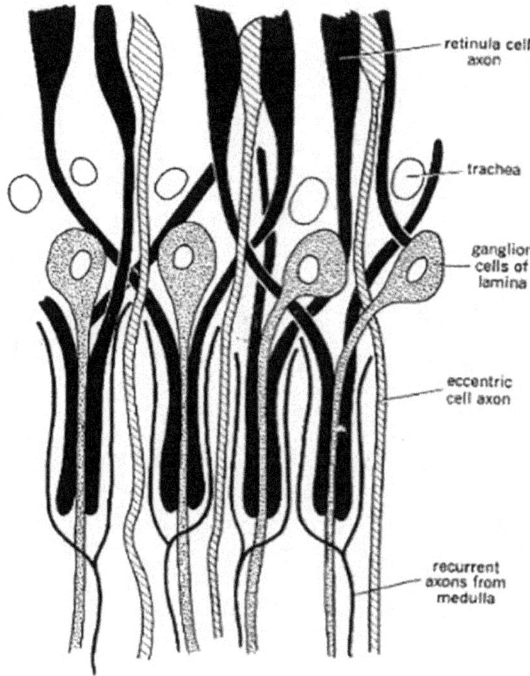

Figure 3. A schematic of the fly lamina, with receptor endings (in black) with lateral shifts to reform the projection from the retina, and the elongated synapses on the lamina cells (grey), and thin fibres that control each synapse.

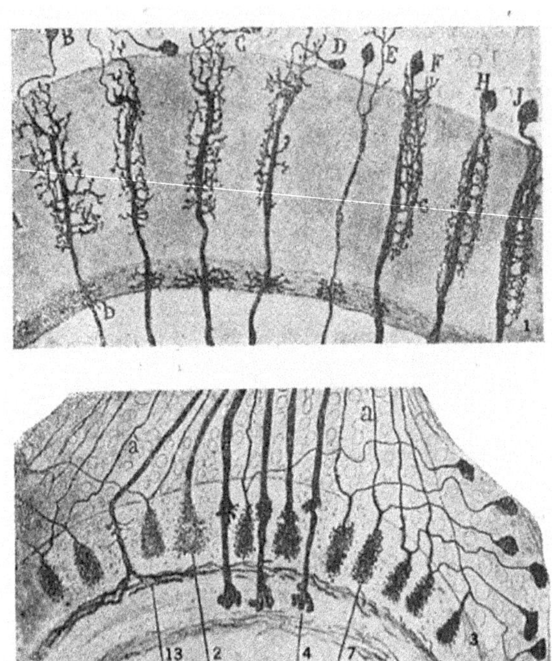

Figure 4. The thick neurons with duplicated synapses in the lamina and huge bushy terminals in the medulla, in silver stained preparations, From the famous works of Ramon y Cajal (1909-1918).

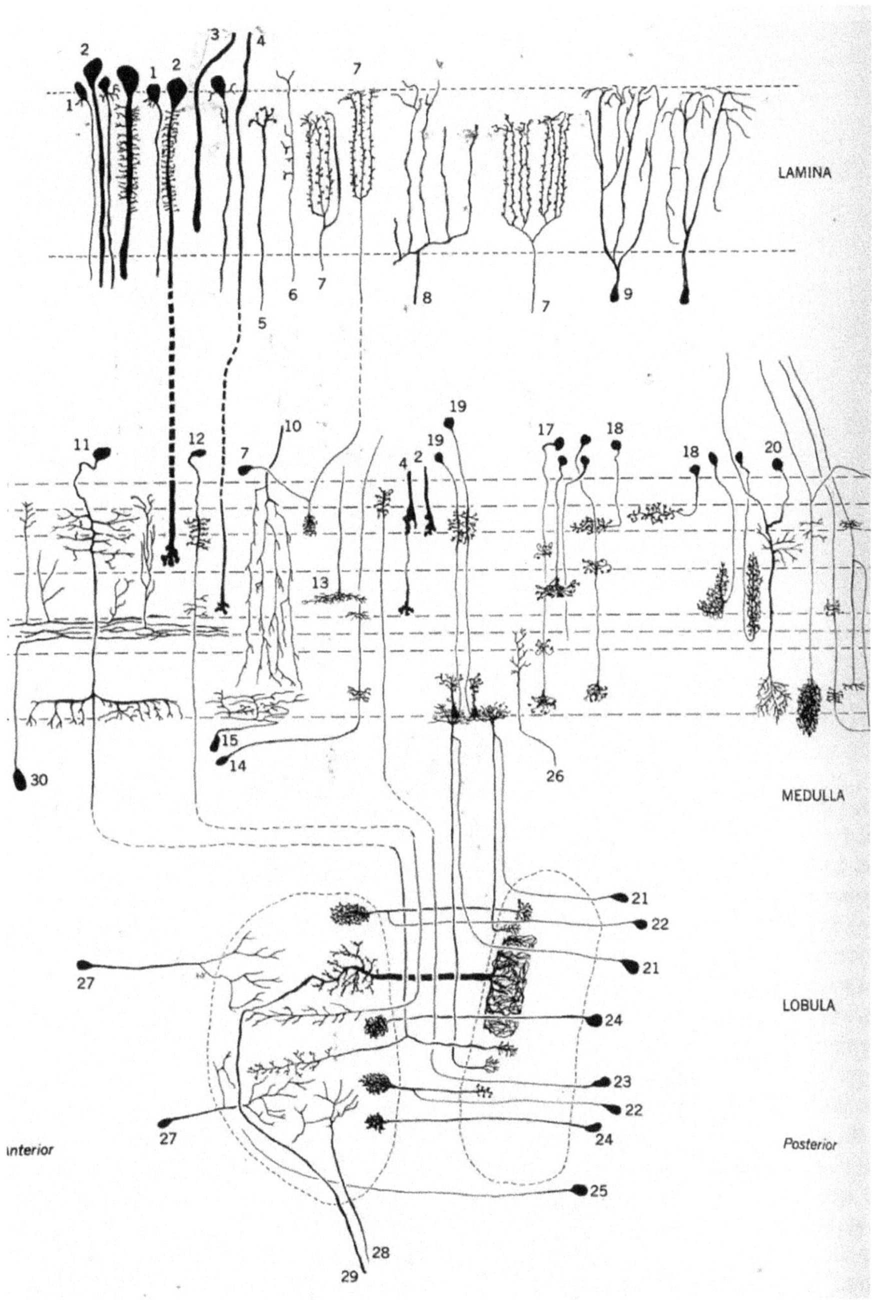

Figure 5. A selection of neuron types of neurons in the three regions of the fly lamina, medulla and lobula of a fly optic lobe. The thick fibres, number 2, are those also illustrated in figure 4. Number 29, which runs to the ventral cord and thorax, is probably also involved in the escape response. (Page 1080 in the book by Bullock and Horridge, 1965).

A Boost for the Green Receptors

Visual cells 1-6 in all flies contain an additional visual pigment with peak sensitivity in the ultraviolet (Figure 6). This increases the strength of the signal that is passed on to the escape response, but has an unfortunate effect on the perception of movement in the visual field.

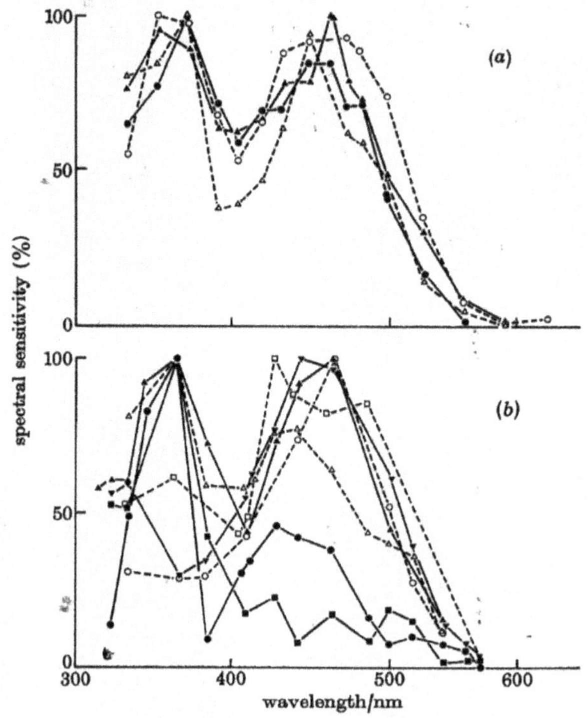

Figure 6. Data from recording the spectral sensitivity of two single cells 1-6 in the retina of *Eristalis*, showing the peaks of two separate visual pigments. The booster near 350nm and blue-green near 450nm (Tsukahara and Horridge, 1977a).

Detection of Motion and Colour

There is a general principle that the detection of motion is always done by detection of a shift in green contrast. First, green light makes up the largest component of sunlight and is the usual colour of natural surroundings because photosynthesis uses the optimum bandwidth. Motion detection in all animals uses the motion of edges, especially between shadows and green, where there is a maximum change in intensity. Obviously, if the detector was equally sensitive to all colours, a movement would be less noticeable. So, the narrower the spectral bandwidth of the light used, the sharper is the motion detection. The UV boost in receptors 1-6 (Figures 6 and 7) therefore spoils the detection of motion.

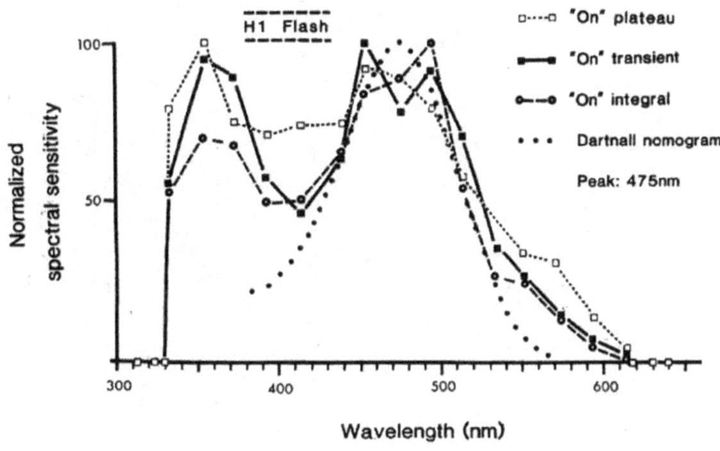

Figure 7. Data from recording the spectral sensitivity of a motion detector neuron H1 of *Eristalis* to a flash stimulus. (Srinivasan and Guy, 1990)

The H1 neuron of *Eristalis* normally detects motion. In recordings when the stimulus is a flash, there is no time to inhibit the effect of the UV input (Figure 7), but the effect of the U boost is eliminated when the stimulus is motion of an edge (Figure 8)

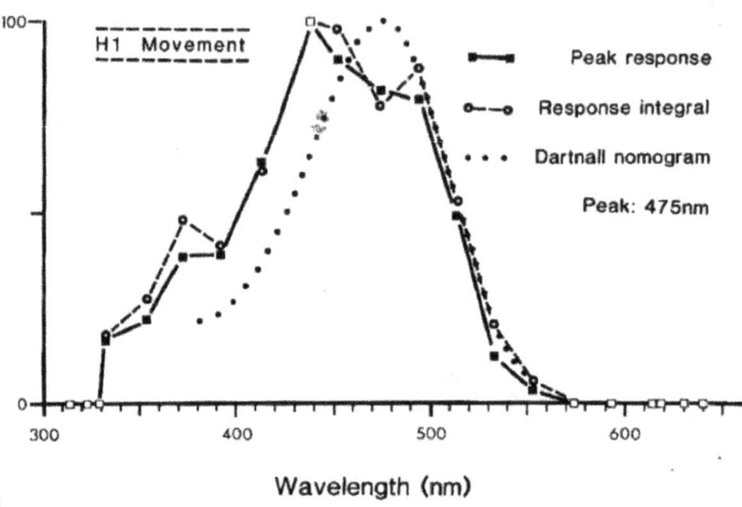

Figure 8. Data from recording the spectral sensitivity of a motion detector neuron H1 of *Eristalis* to a motion stimulus. (Srinivasan and Guy, 1990)

In agreement with the need to distinguish colours, the honeybee has no booster UV pigment in the photoreceptors, so has only a single peak in the spectral sensitivity curve for a flash stimulus, as in Figure 8, whatever the input. Unlike many flies, *Eristalis* and the bee forage among foliage where they have good reason to ignore movements of tonic green, but must detect green contrast and measure the amount of at least one other colour relative to green.

In recordings from large motion-perception neurons of the optic lobe of *Eristalis*, the spectral sensitivity function of the response to brief flashes has two peaks, that correspond to the two peaks of photoreceptors 1-6. However, *Eristalis* responses to motion show only one peak, at about 450 nm, in the blue part of the spectrum, showing that the UV part of response to motion of green edges (the booster) have been suppressed by cell 7y activity, but responses to green contrast remain (Srinivasan and Guy, 1990). Unlike typical flies, *Eristalis* and the bee forage among foliage where they have evolved to detect the movements of green edges and they measure amount of blue. *Eristalis* has adapted its visual detectors to resemble those of the honeybee.

Retinal Receptor Cells 7 and 8 of Typical Flies.

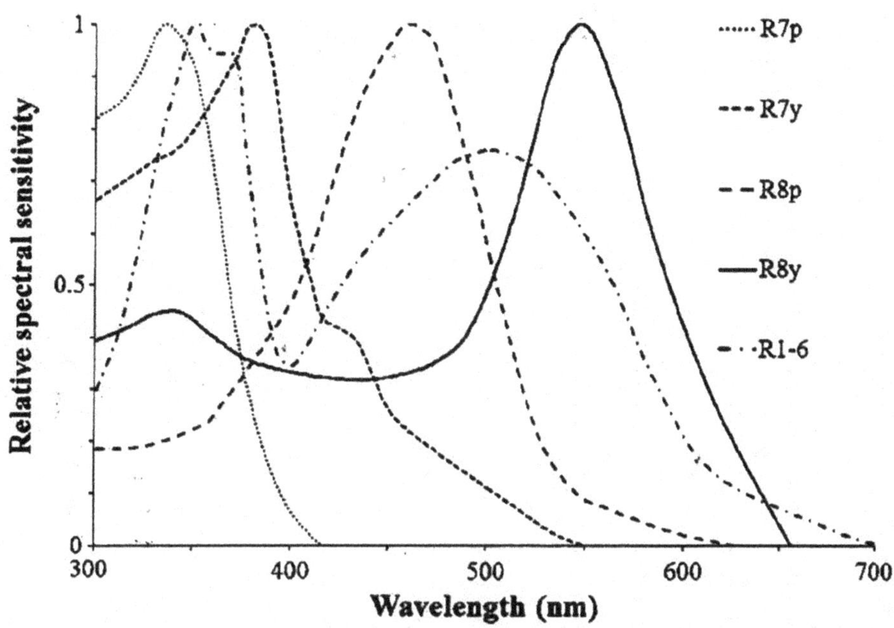

Figure 9. Normalized spectral sensitivity of the, 5 types of photoreceptor cells of the blowfly, *Lucilia*. From left to right, the peaks are R7p, R1-6 near 360nm, R7y, R8p, R1-6 near 500nm, R8y.

Electrophysiological recording shows that in the ommatidia of the blowfly *Lucilia*, a typical fly, there is a bewildering variety of cell types other than cells 1-6 (Tsukahara and Horridge, 1977; Franceschini, N. 1984, Hardie, 1985). The commonest type, Cell 7y has a single peak near 350nm sensitive to UV, and the other, Cell 7p, has a peak near 400nm, sensitive to blue. Their axons run through the lamina to the medulla (See Figure 5).

There are one of two types of Cell 8, one Cell 8p, with peak sensitivity near 450nm in the blue-green, the other, Cell 8y, with a peak near 550nm in the yellow. Cell 8y has a peak at 550nm, in the green yellow part of the spectrum.

At the front of the eye of the male, there is a region looking forward with a large binocular overlap (Figure 9B). Here large Cells R7r (red) have a remarkable sensitivity from 300 to 600 nm. This wide spectral sensitivity is used when chasing a female because it is good for detecting and keeping in view a black spot against a variously coloured background. Unlike normal Cell 7, the axons stop in the lamina, but these cells drive a separate group of neurons in the third neuropile, the lobula (See Figure 5).

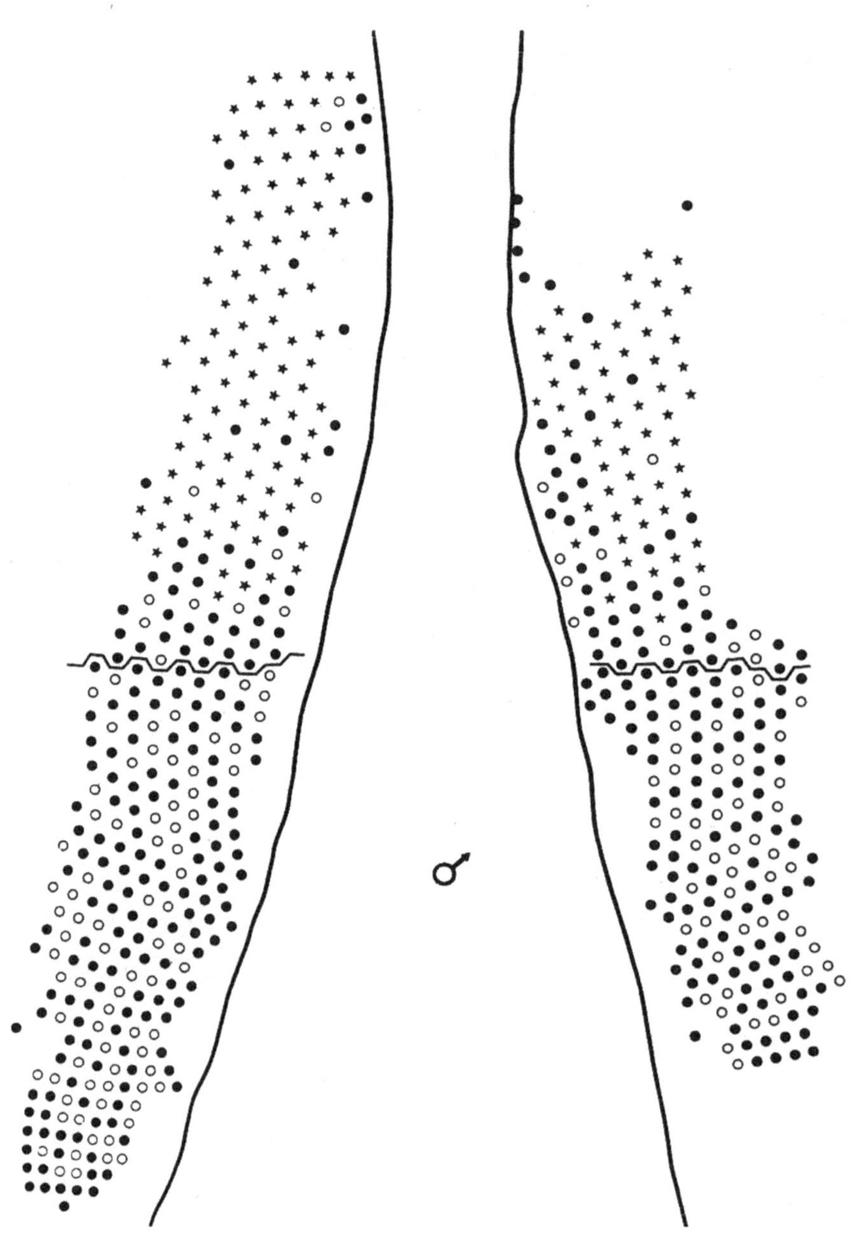

Figure 9B

Colour Vision of Flies

We have the same problems with fly colour vision as we had previously with the honeybee. One is the belief that when a fly is trained to come to a reward of coloured sugar, it must be able to see the colour. Also, because a fly has three or more types of colour receptor behind each facet of the eye, it must distinguish colours in much the same way as humans (Horridge, 2019). Another is that bees also see grey and respond to black, although there are no grey or black photons (Horridge, 2023).

A recent view is that in flies, retinula cells 1-6 in every ommatidium detect motion (which is true) and green (which is incorrect), and cells 7-8 make a colour vision system, especially the many species of flies that are coloured or visit flowers for food. In addition, Lunau (2014) says that both R7 and R8 retinula cells in a single ommatidium contact second order cells in the medulla, which demonstrates that they have colour vision. That is nonsense.

We already know that **bees are dichromats**. In each ommatidium they have (1) six green receptors that detect green contrast but not green as a colour, (2) only one blue receptor in each ommatidium that detects the amount of blue in the visual field of that ommatidium relative to the green background, and therefore its average place in the spectrum; i.e., its colour. It is important to understand that having many ommatidia, this is sufficient visual input to distinguish colours, as indeed we observe in the bee and in most mammals other than primates. (3) As in all animals that fly, bees also have receptors of UV, that inhibit vision of pattern and colour, but show the direction of "up", and the escape route to the sky, and, in dorsal rim facets, information about compass directions.

Consider the cell 8y with peak sensitivity in the yellow. Copying the conclusion about blue in the bee, we can infer that *Eristalis tenax* detects colours as being *more yellow or less yellow* than the green background. They do not merely detect yellow. This agrees well with their food supply and observed behaviour (Figure 10).

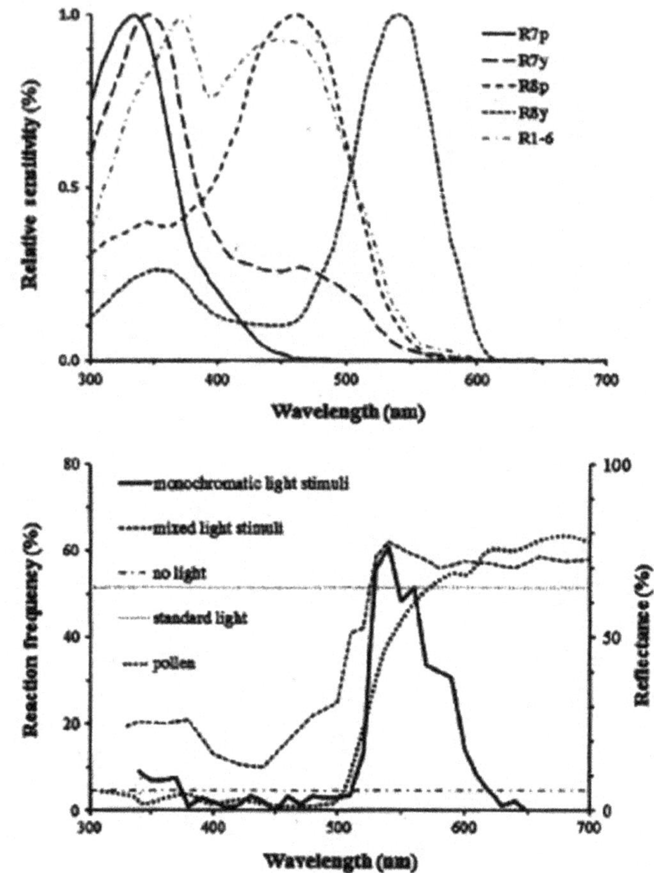

Figure 10. Upper figure. Spectral sensitivity of the photoreceptor types of *Eristalis tenax*, and of ambient light and pollen, abstracted from Horridge et al, (1975) and Tsukahara and Horridge (1977a, b). Lower figure. Innate preference of the proboscis reflex of *Eristalis* to monochromatic light stimuli (solid line) and mixed light stimuli (dotted lines). Showing cell 8p preference at 550nm, far from the range of cells 7 or1-6.

The First Synapse and the Lamina

The axon from cell 7, and its mate from cell 8 in the same ommatidium, pass directly through the lamina and terminate in the next neuropile region, the medulla. In the 1970's, researchers in Tubingen showed that each cartridge of lamina ganglion cells receives six cell 6 receptor axons that look in the same direction, so the lamina has the same spatial layout as the retina. In the 1970's, students in my laboratory showed that the lamina ganglion cells detect *the rate of change* in the green receptor terminals looking in one direction, so they detect and measure green contrast, and record its position. They do not see green as a colour of areas because flies evolved in a world that consists mainly of boring areas of green chlorophyl that is of no interest, but contrasts at edges of shadows and motions of edges, are of great interest to a fly.

The axons from each lamina cartridge (second order cells) terminate in the next ganglion, the medulla, forming again an array that replicates the layout across the retina. This time the array includes the terminals of cells 7 and 8, so that the spatial layout now has information about three different parts of the spectrum.

We can now refer to the bee, which has exactly this neural geometry at this level, except that the bee has green contrast, an array that shows the vertical from the distribution of UV, and an input that indicates more or less blue than there is green in view. The fly has a strong input that shows the distribution and motion of green contrast, also an input from UV that shows the direction of the sky for escape, and an array that shows an indication of colour by measuring the amount of yellow relative to green at each place in the array. The bee operates in a world that often includes blue flowers, whereas the fly is more interested in a world that includes blood and meat. Each insect sees best what is related to food.

However, "what flies see" depends on the properties of the next stage in vision, the feature detectors, about which we know nothing because it is a tedious task to train a fly and then test it with variety of possible cues. Despite that, we know a great deal about the fly eye detects, and what is processed in the optic lobes, as revealed by recording with micro-electrodes in the optic lobes and brain,

Vision of Grey

There re no grey photons, but insects of all kinds distinguish levels of grey by its different content of white, which has a particular level of blue for bees or yellow for flies. Therefore, Cells7 or 8 manage this task. Black/white photographic film records grey levels by the different percentage of silver grains. Humans detect grey similarly by the percentage of retinal rods that generate black when they lack a photon stimulation (Horridge, 2023).

When humans seeing grey concentrate on what they actually detect with their eyes, they see millions of tiny black dots that are in their own eyes Each dot is generated by a retinal rod that lacks a photon. So, the fewer the photons, the more dots there are. In normal vision, as we glance at it, we hallucinate the grey level corresponding to the density of dots. Humans never actually see grey levels because there are no grey photons.

In Conclusion; What Principles Have We Learned?

My aim has been to infer what insects see and how they achieve vision without a large brain or multiple layers of processing, with an intention to build later such systems with silicon chips. Because *bees* learn a target and return to it, they can be trained to and tested exhaustively until we infer what they really detect and remember. The main result was that they detect in all directions simultaneously a flow of edges of green contrast from edges. They detect discontinuities that indicate parallax, from which they infer local range in each direction, and prevent collisions, but they do not infer closed circuits of edges or shapes. In area of colour between lines of contrast, bees also detect the amount of blue, which give s them an indication of its position in the colour scale, whether more blue or less blue than green background. Therefore they are dichromats, like of most grazing animals, cats and dogs. UV provides flight stability, as in birds and surface fishes. The UV receptor gives them the direction of the vertical to make steady flight possible. Flies operate in a similar way as dichromats, but use a yellow sensitive Cell 7y in a similar way, recording amount of yellow relative to the green background. Having poor memory, flies are very difficult to train, but there are enough clues to demonstrate that they have some colour discrimination (Ilse, 1949; Lunau, 2014).

References

Bullock, T. H. and Horridge, G. A. 1965. Structure and Function in the Nervous Systems of the Invertebrates. San Francisco; Freeman and Co.

Franceschini, N. 1984. Chromatic organization and sexual dimorphism of the fly retinal mosaic. In "Photoreceptors" Ed. E. Borsellino and L. Cervetto. Plenum Press.

Golding Y. C. and Edmunds M. 2000. *Behavioural mimicry of honeybees (Apis mellifera*) by droneflies (Diptera: Syrphidae: *Eristalis* spp.). Proc. R. Soc. Lond. B 267, 903-909: doi: 10.1098/rspb.2000.1088

Hardie R. C. 1979 Electrophysiological analysis of fly retina. I. Comparative properties of R1-6 and R7 and R8. J. Comp Physiol. 129, 19-33.

Hardie R. C. 1985 Functional organization of the fly retina. In: Ottoson D. (Ed.) Progtress in Sensory Physiology. Vol 5. Springer 1979.

Horridge, G. A. 2019 The Discovery of a Visual System; The Honeybee. CABI Books. Wallingford and Boston. Also ePub. ISBN-13:978 1 78924 091 7

Horridge G. A. 2023, *How do bees (and humans) see grey levels*? Northern Bee Books, Hebden Bridge. www.northernbeebooks.co.uk

Horridge G. A. Mimura, K and Tsukahara Y. 1975, Fly photoreceptors. II Spectral and light sensitivity in the drone fly *Eristalis*. Proc. R. Soc. Lond. B 190, 225-237.

Ilse, D. 1949 Colour discrimination in the drone fly, *Eristalis tenax* Nature *London,* 163, 255-256.

Lunau K. 2014. Visual ecology of flies with particular reference to colour vision and colour preferences. Journal of Comparative Physiology, 200, 497-512. DOI:-10.1007/s00359-014-0895-1

Srinivasan M. V. and Guy, R. G. (1990) Spectral properties of motion perception in the dronefly *Eristalis*. J. Comp. Physiol. A 166, 287-295

Tsukahara Y. and Horridge GA.1977a. Visual pigment spectra from sensitivity measurements after chromatic adaptation of single dronefly retinula cells. J. Comp. Physiol. 114; 233-251.

Tsukahara Y. and Horridge GA.1977b Interaction between two retinula cell types in the anterior eye of the dronefly *Eristalis*. J, Comp. Physiol. 115, 287-298.

www.ingramcontent.com/pod-product-compliance
Lightning Source LLC
LaVergne TN
LVHW060635110826
845147LV00014B/912
* 9 7 8 1 9 1 4 9 3 4 7 5 9 *